THE ENERGY DILEMMA
UNDERSTANDING GLOBAL ISSUES

Published by Smart Apple Media
1980 Lookout Drive
North Mankato, Minnesota 56003
USA

This book is based on *The Energy Dilemma: The Dominance of Fossil Fuels*
Copyright ©1998 Understanding Global Issues Ltd., Cheltenham, England

Library of Congress Cataloging-in-Publication Data

The energy dilemma / edited by Celeste Peters.
    p. cm.--(Understanding global issues)
Includes index.
Summary: Explores the energy crisis on a global scale, covering the environmental,
political, social, and economic implications of dependence on non-renewable
energy sources.
  ISBN 1-58340-169-5 (hardcover: alk. paper)
 1. Energy development--Environmental aspects--Juvenile literature. 2. Energy
industries--Political aspects--Juvenile literature. [1. Energy development. 2. Energy
industries. 3. Power resources.] I. Peters, Celeste A. (Celeste Andra), 1953– II. Series.
  TD195.E49 E533 2002
  333.79--dc21

                           2001008448
                    Printed in Malaysia
                    2 4 6 8 9 7 5 3 1

EDITOR Jared Keen   **COPY EDITOR** Heather Kissock
TEXT ADAPTATION Celeste Peters   **DESIGNER** Terry Paulhus
PHOTO RESEARCHER Tina Schwartzenberger

Contents

Introduction   5

Energy Revolutions   6

Old and New King Coal   10

Oil: Almost Perfect Energy   17

The Age of Natural Gas   22

Mapping Energy Resources   26

Charting the World's Energy   28

Fossil Fuel Technology   30

Fossil Fuels Forever?   36

Security and Energy   42

Time Line of Events   46

Concept Web   48

Quiz   50

Internet Resources   52

Further Reading   53

Glossary   54

Index   55

Introduction

Our world economy is based on **fossil fuels**. The factories that make products and the vehicles that transport the products to market are powered by energy from fossil fuels. Coal alone generates more than one-third of the world's electricity.

For years, people have claimed that this fossil fuel economy is doomed. Yet coal, oil, and gas still produce most of the energy sold worldwide. In fact, the world depends on fossil fuels more every day. There is still no other practical way to meet the world's energy demands. During the next 50 years, sales of **renewable energies**, such as wind power and solar power, could grow greatly. At present, though, fossil fuels rule.

Coal is the main energy source in China and India, where a large portion of the world's population lives. Modern industry is spreading throughout these two countries. As this happens, the demand for fossil fuels will grow. This could greatly drain world fuel supplies. It could also increase pollution, including levels of greenhouse gases.

The amount of fossil fuels burned to produce energy has nearly doubled every 20 years since 1900.

Many people link the use of fossil fuels with climate change. They argue that we must reduce the large amounts of greenhouse gases that the burning of fossil fuels releases into the air. Otherwise, the average world

We need to minimize the health and environmental costs of energy production.

temperature could increase to a level that would cause significant environmental damage.

The energy dilemma is clear. We need to minimize the health and environmental costs of energy production. How do we do this without adversely affecting the world economy?

Several countries have made a formal promise to cut greenhouse gas **emissions**. There is also room to improve energy efficiency. Current technologies are very wasteful. For example, only 20 percent of the energy in gasoline powers a car. The rest is lost as heat.

Renewable energy sources could be another solution. Several "green" fuels (those made from plants) are being developed. A vast amount of energy is also available in sunshine, wind, and other renewable sources, but these energy sources are not yet concentrated enough to be useful in industry and transportation. While technology wrestles with the energy dilemma, we continue to produce and use fossil fuels in huge amounts. The world uses 3.3 billion tons (3 billion t) of oil per year and similarly large amounts of coal and natural gas. This consumption cannot continue forever. Eventually, we will use all of the available fossil fuels. As such, there is an urgent need for workable energy alternatives.

ENERGY ALTERNATIVES

Major alternative energy sources include:

Hydroelectric—harnessing moving water to produce electricity is a major industry

Geothermal—heat contained deep beneath the ground is a valuable source of energy

Solar—the Sun's rays can provide an almost unlimited amount of energy

Wind—wind farms use the power of moving air to create electricity

Energy Revolutions

Coal powered the Industrial Revolution of the 19th century. Oil ruled the 20th century. Gas could be the front-runner in the 21st century. Today, these three fossil fuels provide most of the energy that drives industrial society.

For most of history, muscles provided the main power source for many tasks. Farming, construction, and manufacturing, for example, all relied on muscle power.

Occasionally, people used a waterwheel or a windmill to lift water or grind grain. These

Fossil fuels provide most of the energy that drives industrial society.

tools were useless, though, when water did not flow or wind did not blow. Muscles were the only reliable power source.

The invention of the steam engine changed all that. The steam engine made it possible for mechanical power to do the work of many people. Coal, oil, and gas became the fuels that powered the new engines. In the industrialized world, machines largely replaced manual labor. In developed countries today, human muscle power is used mainly for leisure activities, such as sports, rather than to lift, dig, or build.

Fossil fuel energy has given a significant boost to human productivity. Take a small lawnmower, for example. It can mow an area of grass faster and cheaper than a team of four horses can. Fossil fuels have also made it easier for us to do household chores. The fuels generate the electricity that powers electrical appliances. These devices allow us to do a wide range of jobs that once required a great deal of manual labor.

Today, we take energy for granted. We seldom think about the effort needed to bring electricity to our houses or gasoline to our cars. Sometimes the power goes off for a while. This reminds us of how much

The waterwheel was a major source of power until the steam engine was developed in the 1700s.

we depend on an unbroken flow of energy. Only a thin line divides our modern lifestyle from the way people lived in pre-industrial times.

In the 1970s, people often claimed that the world would run out of oil in the next 30 years. Clearly, that has not yet happened. Since the 1970s, we have continued to discover oil deposits. In fact, oil companies have been finding **proved recoverable reserves** even faster than we have been using oil.

One of the interesting things about the fossil fuel industry is that proved reserves are greater now than they were 20 years ago. This is the case despite the rapid rate at which the world is using fossil fuels. **Petroleum**

companies continue to discover new fossil fuel deposits and to find better ways to retrieve them.

There is far more oil in the ground than is contained in proved reserves. For example, world deposits of **oil shale**

"Horsehead" pumpjacks first appeared in the 1940s. Although less common today, they can still be found pumping oil from the ground using rotation and vacuum principles.

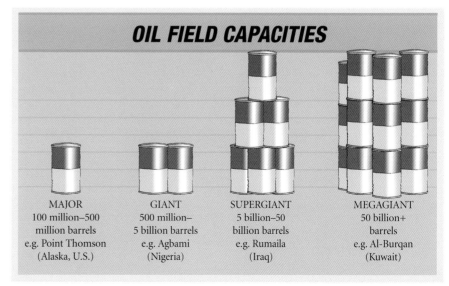

OIL FIELD CAPACITIES

MAJOR	GIANT	SUPERGIANT	MEGAGIANT
100 million–500 million barrels	500 million– 5 billion barrels	5 billion–50 billion barrels	50 billion+ barrels
e.g. Point Thomson (Alaska, U.S.)	e.g. Agbami (Nigeria)	e.g. Rumaila (Iraq)	e.g. Al-Burqan (Kuwait)

HOW LONG WILL IT LAST?

In 1998, studies revealed that the world's proved oil reserves would last for nearly 41 years if we kept producing oil at the same rate as we did in 1997. Production of world oil, gas, and coal were at all-time highs that year. In the future, production could slow down. People in developed countries are becoming more concerned about the environment and long-term oil supplies. This means that we are likely to rely more and more on renewable forms of energy and nuclear power within the next few decades. As a result, the world's proved oil reserves could last for longer than 41 years, which is good news, considering the fact that oil is used in plastics and has many other non-energy uses.

There is growing unease about our fossil fuel economy. It is dangerous to rely so heavily on non-renewable fuel sources. In spite of this, our dependence on them is increasing instead of decreasing. Rapid economic development in Asia and Latin America has had much to do with this. At the same time, studies have linked fossil fuels to pollution and global warming. A move to cleaner, renewable energy is urgently needed.

contain huge amounts of oil. We cannot recover this oil at a profit using current technology. In fact, we may never recover it, even if our technology improves. People are increasingly concerned about harming the environment. Also, the technology to tap into renewable energy sources is improving. Oil shales, therefore, seem likely to stay in the ground. So, too, does the oil contained in **tar sands**.

In the industrialized world, fossil fuels account for 89

Tankers haul nearly all the oil imported by the United States. The largest tankers can hold more than one million barrels of petroleum.

percent of the total primary energy supply. Oil alone accounts for almost 40 percent. Oil is strongly linked with the Middle East. Desert oil wells there contain nearly two-thirds of the world's known reserves.

The development of oil and gas resources requires a large amount of money. As a result, only a handful of wealthy companies control the global oil industry. The largest oil companies are powerful players in the global economy. They have offices and oil fields all around the world. Lately, though, they have run into increased competition. New oil companies

Most commercial wind turbines are capable of generating 500–705 kilowatts per hour.

have come onto the scene in the developing world. Also, large Russian companies such as Gazprom and Lukoil have entered the market.

More than 41,000 oil fields have been discovered so far. They are ranked according to size. "Megagiants" are the largest oil fields. The world's two megagiants are Al-Ghawar in Saudi Arabia and Al-Burqan in Kuwait. Al-Ghawar contains 86 billion barrels of oil and Al-Burqan contains 75 billion barrels. So far, we have discovered about 40 "supergiants." Most of these are located in the Middle East. There are also about 300 "giants" and 1,000 "majors." Megagiants, supergiants, giants, and majors account for most of the world's recoverable oil. The smallest oil fields are "tiny" and "insignificant."

KEY CONCEPTS

Global warming Studies have shown that the average global temperature is increasing. A change of only a few degrees can greatly alter climates. While debate continues over the exact cause of global warming, air pollution is thought to play a key role.

Primary energy supply The energy obtained from all sources is a country's primary energy supply. The primary energy supply includes imports, too, but not exports.

Renewable and non-renewable sources Unlike fossil fuels, some sources of energy are potentially endless. The Sun, wind, and water are renewable energy sources. No matter how much of these sources we use, we can always obtain more. On the other hand, fossil fuels exist in limited amounts. They are non-renewable energy sources and will eventually run out or become economically unviable.

Old and New King Coal

The Chinese have been mining coal for more than 2,000 years. Europeans only began using coal widely in the 18th century. Coal became the main energy source in the parts of the world that embraced the Industrial Revolution. It replaced wood as the fuel of choice to drive steam engines. Coal oil and "town gas" made from coal replaced whale oil as the main fuels for lighting. Coal fires were used to heat households because coal was more efficient and less bulky than wood. Coal mining became a major industry in Europe and North America. It employed millions of people, many of them children. Today, we use coal in the West mainly to generate electricity in large power stations. In other parts of the world, coal is used for a variety of purposes. Countries such as China and India still use coal as the main energy source for houses and industry.

Coal comes in many varieties. In general, there

are four main coal groups, based on carbon content and how much heat each type of coal produces when burned. The "soft coals" burn at relatively low temperatures. The soft coals are **lignite**, or brown coal, and **sub-bituminous coal**. The "hard coals" burn at high temperatures. Hard coals are **bituminous coal** and **anthracite**. Anthracite burns hot and has low emissions of soot and sulfur. It is regarded as the premium variety among all the coals.

Different coals are used for different purposes. Most types, though, can be used to generate electricity. Coal that burns at a high temperature is known as **metallurgical** coal. It is used to make steel and other metals. Three-quarters of the world's steel is made using **coke**.

Coal deposits exist all around the world. The biggest known reserves are in Australia, China, Germany, India, Poland, Russia, South Africa, and the United States. Not all countries have the same types of coal. For example, China has mostly bituminous coals. Germany and Poland have mostly soft, high-sulfur coals, such as lignite. The United Kingdom and eastern United States have large deposits of anthracite.

More than half of the coal used comes from underground mines. The rest is taken from

> It is estimated that the world's supply of coal can last 250 years, if the present rate of use remains the same.

surface deposits. In developed countries, much of the mining is done using computer-controlled cutting machines. This method has largely replaced labor-intensive coal mining. Open strip mining can produce even higher yields than underground mines can, but this method has a devastating impact on the environment, as vegetation and soil are removed with entire hillsides. Large open strip mines are common in the United States, Canada, Australia, South Africa, and Indonesia.

Mines that require a great deal of human labor remain common in less-developed countries. Tunnel cave-ins and gas explosions are problems in deep mines. Mining accidents are responsible for more than 5,000 deaths in China each year. By comparison, 29 workers died in U.S. mines in 1999.

In the West, coal mines employ far fewer workers today than they once did. They also use modern management skills and advanced technology. Over time,

MAJOR WORLD COAL PRODUCERS

China	1,290 million tons (1,171 million t)
United States	980 million tons (889 million t)
India	341 million tons (310 million t)
Australia	262 million tons (238 million t)
South Africa	248 million tons (225 million t)
Russia	186 million tons (169 million t)
Poland	112 million tons (102 million t)
Ukraine	89 million tons (81 million t)
Indonesia	87 million tons (79 million t)
Kazakhstan	78 million tons (71 million t)

COAL CONSUMPTION

In terms of overall use, these countries were the top consumers of coal in 2000:

1. United States — 26 percent of the world total, equivalent to 621.8 million tons (564.1 million t) of oil

2. China — 22 percent of the world total, equivalent to 529.2 million tons (480.1 million t) of oil

3. India — 7 percent of the world total, equivalent to 179.8 million tons (163.1 million t) of oil

4. Russia — 5 percent of the world total, equivalent to 121.7 million tons (110.4 million t) of oil

the move to these practices has made quite a difference. It has reduced the harm that mining inflicts on the health of workers and the environment. The savings, however, come at a price. The changes have destroyed entire communities whose way of life relied on coal mining for more than 100 years. This shift took place in many developed countries during the past century. The United States, for example, had more than 700,000 coal miners in 1923. This number fell to roughly 70,000 by the year 2000. On the other hand, some towns in Russia continue to exist today solely due to nearby mines.

Many giant companies run the oil and gas industries from offices all around the world. Only a few giant companies run the coal industry. The United States and Australia have large private coal companies. In most other parts of the world, state-owned coal businesses, such as Coal India in India, are the rule.

The United States gets more than half of its electricity from coal-fired power stations. Most of the coal produced in the country is used for this purpose. Mines in Wyoming, West Virginia, Kentucky, and other states fill the demand.

Coal gives off a large amount of heat when it burns. This makes it a valuable fuel. Coal's main drawback is the pollution it causes. Mining, transporting, processing, and burning coal are all activities that contribute to pollution and, possibly, global warming. Sulfur and ash are two

The U.S. gets more than half of its electricity from coal-fired power stations.

of the pollutants released when burning coal. Soft coals contain especially high amounts of these two substances. All coals give off greenhouse gases, such as carbon dioxide. They may also emit tiny

radioactive particles when they burn. In fact, some people claim that radioactive fallout from coal-fired power plants is higher than fallout from nuclear power plants.

New technology can change coal into a relatively "clean" fuel to burn. The cost is high, though. Washing coal reduces its ash content but creates its own set of problems—it uses large amounts of energy and creates wastewater. Removing pollutants as the gas leaves the smokestack is also expensive. On a worldwide basis, this is seldom done. There is also no way to transport coal without creating

Contour mining occurs on hilly terrain. Workers remove overlying rock before mining the coal.

pollution. Despite these problems, China, India, and other developing countries base their energy futures on coal.

Coal mining is one of the most dangerous and environmentally damaging of all major industries. True, safety standards have greatly improved, but mining accidents still kill thousands of miners. Many miners are also harmed by breathing coal dust and working in dark, wet, and cramped conditions. Yet society tolerates coal mining.

Coal usage is rising. In fact, the United States Department of Energy expects world coal usage to increase by almost 50 percent during the next 20 years. China

KEY CONCEPTS

Greenhouse gases The six main ones are carbon dioxide, methane, nitrous oxide, hydrofluorocarbons, perfluorocarbons, and sulfur haxafluoride.

Kyoto Protocol This agreement was formally adopted by 84 countries in Kyoto, Japan, in 1997. It requires industrialized countries to reduce their greenhouse gas emissions by 2012. This agreement has not yet been ratified in the U.S. In 2001, the U.S. renounced the treaty on the grounds of cost.

Open strip mining This method of mining consists of obtaining coal from deposits located close to the ground's surface. In open strip mining, large machines pull away or "strip" coal from exposed seams.

and India will account for most of the growth. Asian coal usage is expected to double between 2002 and 2020. Asian imports of coal are predicted to rise by more than 70 percent during this same time period. This will bring the amount of coal Asia imports to 485 million tons (441 million t).

Much of this coal will come from open-pit mines in Australia and South Africa.

These forecasts may have to be revised. Asia currently has a financial crisis on its hands. As a result, it may not be able to afford the predicted levels of imports. Also, the Kyoto Protocol,

The world contains more than 1.1 trillion tons (1 trillion t) of proved recoverable coal reserves.

an international treaty, could affect coal usage. Yet there seems little doubt that coal will play a key role in the world's energy supply for many more years.

Duties: Maintain and oversee the operations of power plants
Education: A certificate or diploma in power engineering
Interests: Engineering, building operations, and energy production

Navigate to the Petroleum Institute Web site: 212.78.70.142/index.cfm?%20PageID=48 for information on related careers.

Careers in Focus

Power-plant operators control the machinery that generates electricity. Operators monitor boilers, **turbines**, and generators, as well as instruments that control voltage and regulate the flow of electricity from the plant. If local or national power requirements change, plant operators are poised to switch generators on or off and change circuit connections. They use computers to keep track of switching operations and power loads on generators, lines, and transformers.

Some electric power-generating plants are automated. Operators in these plants work in a central control room and are often called control-room operators. In older plants, the equipment controls are less centralized.

Often, workers are assigned rotating shifts because many power plants operate 24 hours per day. This can be stressful and tiring because sleeping patterns are constantly changing. Operators who work in control rooms generally sit or stand at a control station. The work is not physically strenuous but requires close attention. Operators who work outside the control room may be exposed to danger from electric shock, falls, and burns.

About 92 percent of all electric power-generating plant operators work for utility companies and government agencies that produce electricity. The remainder work for manufacturing establishments that produce their own electricity.

Oil: Almost Perfect Energy

Oil provides energy in a cheap, compact form that is easy to use, store, and move from place to place. That is its beauty. Nothing else compares to oil in terms of its range of practical uses. Almost all of the world's aircraft, ships, and motor vehicles run on petroleum fuels. In the developing world, oil still fuels many power stations. Even products that we use daily are made from petroleum-based raw materials. Among them are textiles, plastics, fertilizers, and medicinal drugs.

In the ancient world, people collected oil that seeped from the ground. They used this oil for a variety of purposes, such as the making of flame-throwing weapons. Edwin L. Drake drilled the first U.S. oil well in 1859 in northwestern Pennsylvania. Back then, oil was the fuel used in lamps. Today, oil still provides us with light but not directly. It fuels the generators that supply the electricity to light bulbs. Oil is just one of several fuels that can run generators. Coal, gas, nuclear energy, or any of the renewable energy sources will work, too. Still, nothing can compete with oil when it comes

The average land rig costs about one-tenth as much to build and operate as an offshore rig does.

to powering motor vehicles. The invention of cars has created a second and even larger market for oil than lamps did. Light **crude oils** are just right for the production of motor fuel. These oils typically come from Saudi Arabia and other Middle Eastern countries.

Nothing can compete with oil when it comes to powering motor vehicles.

The world produces and uses more than 3.3 billion tons (3 billion t) of oil every year. This amounts to about 70 million barrels a day. More than half of this oil sells either as crude oil or as **refined** petroleum products. About 2.2 billion tons (2 billion t) of oil are traded between different countries each year.

The storage and distribution of crude oil and petroleum products is a major industry. The same companies that look for and produce the oil often control this part of the business. These large firms do much of the crude oil refining. They also produce many oil-based chemicals and fuels.

Pipelines usually move oil from **wellheads** to refineries. Often, the refineries are located near seaports. This makes it

easier to ship refined products by tanker. Nearly 8,000 oil tankers sail the world's oceans. A large number of these weigh more than 275,000 tons (250,000 t) when they are full. About half of all tankers are registered in Liberia, Panama, Greece, or the Bahamas.

Chartered ships often carry the oil. This raises concern over the variety of standards used, although international laws are much tougher than they used to be. Most large companies use double-hulled tankers and enforce strict pollution-control regulations. Some smaller companies apply looser rules, increasing the risk of spills at sea. Even so, the amount of oil that escapes into the world's oceans is tiny when compared to the quantities carried back and forth. In any case, oil is a natural product that eventually breaks

OFFSHORE OIL

Offshore exploration is gaining in popularity. It began in the Gulf of Mexico and has spread to many other parts of the world. Wells can now be drilled farther than 200 miles (300 km) offshore in water more than 6,560 feet (2,000 m) deep. Offshore exploration occurs in the waters of about half of all nations.

THE OIL COMPANIES

Most of the leading U.S. oil companies had their origins in the Standard Oil Company founded by John D. Rockefeller in 1870. Shell arose from a 1907 combination of Dutch and English companies. British Petroleum (BP) was founded as the Anglo-Persian Oil Company in 1908. Today the largest majors are Exxon-Mobil, Shell, BP-Amoco, and Chevron-Texaco.

Other international oil companies include ENI (Italy); Elf Aquitaine and Total (France); Atlantic Richfield and Phillips Petroleum (U.S.); and Statoil (Norway).

Major Middle Eastern enterprises include Aramco (Saudi Arabia) and NIOC (Iran).

Although U.S. and European companies continue to dominate the industry, new competitors have come into the global oil market in recent years. They include Pemex (Mexico); PDVSA (Venezuela); Sunkyong (South Korea); Petrobras (Brazil); Petronas (Malaysia); Lukoil (Russia); Indian Oil; and Chinese National Petroleum.

The world's leading supplier of services and technology to the international petroleum industry is Schlumberger. Founded in 1927 in France, Schlumberger employs 80,000 people worldwide.

down in seawater. Chemicals used to break up the oil may do more harm than the oil itself. On the other hand, heavy oil spills close to shorelines can harm local sea birds and other forms of marine life.

Of even greater concern are the effects of burning oil and other fossil fuels. Burning these releases carbon dioxide, sulfur dioxide, nitrous oxides, and volatile organic compounds (VOCs) into the air. Carbon dioxide is the greenhouse gas most tied to global warming. Sulfur dioxide mixes with water vapor in the air and falls as acid rain. Nitrous oxides lead to the formation of smog. VOCs also

include substances that can be harmful to the health of people and animals. VOCs escape into the air when gasoline is not kept in airtight containers. This allows the gasoline to evaporate. Leaks

Until the 1950s, the U.S. led the world in oil production.

from pipes or tankers that are being loaded or unloaded also add to the buildup of VOCs in the air.

The United States and Western Europe together

account for half of all oil imports. Until the 1950s, the United States actually led the world in oil production. Since then, the Middle East has taken the lead. It now accounts for nearly one-third of world production, two-thirds of proved reserves, and almost half of global exports.

Many Middle Eastern nations belong to OPEC, the Organization of Petroleum Exporting Countries. The original five members were Iran, Iraq, Kuwait, Saudi Arabia, and Venezuela. These were followed by Algeria, Indonesia, Libya, Nigeria, Qatar, and the United Arab Emirates. Ecuador and

In the first nine months of 2001, Colombia produced 618,000 barrels of oil per day.

Gabon were members for a while but dropped out in the 1990s.

OPEC ensures its members' oil policies agree with one another. The 11 OPEC countries produce more than one-third of the world's oil. They contain more than three-quarters of the world's proved oil reserves. The OPEC countries also have nearly all of the world's ability to increase oil production. However, many countries with large oil reserves cannot develop them without help from western countries. They simply lack the money and technology to do so. From Ecuador to Iran, there is fierce competition to attract foreign money.

One of the most active areas for oil and gas deals is the Caspian Sea, in central Asia. This area's large potential reserves have attracted much interest, but the complex politics

THE U.S. OIL INDUSTRY

Edwin Drake drilled the first U.S. oil well in 1859. At 69 feet (21 m) deep, the Pennsylvania well produced 15 barrels per day. Soon, the area boomed, and the U.S. oil industry was born. Huge oil reserves were later discovered in Oklahoma and Texas, which became major centers for U.S. oil production.

The amount of crude oil produced in the U.S. reached its peak in 1970 at 11.3 million barrels per day. In 2000, average per day production was 8.1 million barrels. About 1.4 million people are employed in the U.S. oil industry.

of the region have delayed the carrying out of many huge projects. Iraq, with its giant oil fields, is in the same situation. Iraq is second only to Saudi Arabia in the size of its proved reserves. The United Nations has imposed trade sanctions on Iraq, preventing international trade.

The Middle East offers many advantages as a source of oil. Its crude oil is of a high quality, yet it is inexpensive. It is also relatively easy to extract from the ground. In addition, the oil is located in a central global region that has easy access to sea transportation. There may be plenty of oil elsewhere in the world, but it is more expensive to extract and more problematic to transport.

Oil is the ultimate "risk and reward" business. Most oil wells do not yield enough oil to be profitable. Even with detailed geological maps and a century of experience, the industry has trouble finding oil. On average, 10 test wells produce only 1 profitable well. The rewards often justify the risks, though— the huge Prudhoe Bay field in

the Arctic was found only on the last of a series of test wells.

Governments around the world know the political importance of cheap fuel. They spend more than $100 billion a year to support the electricity and fossil fuel industries, according to the World Bank. If governments were to cut back their support, it would save

Refineries convert petroleum into useful products. Petroleum is separated, converted, and chemically treated.

them money and improve energy efficiency. Unfortunately, it could also lead to a rise in the price of fuel. This would be politically risky. Many people depend on fuel being inexpensive.

KEY CONCEPTS

Offshore exploration Test wells are often drilled into the ocean floor. Oil companies construct platforms above the water from which these wells are dug. Offshore exploration is a costly venture, but large reserves of oil have been found. The Gulf of Mexico and Alaskan waters are two areas where extensive offshore exploration has occurred.

OPEC The Organization of Petroleum Exporting Countries was formed in Baghdad, Iraq, in 1960. This group of the world's major oil exporters has great influence over the price of oil.

Refined petroleum Crude oil must be processed. The refining process separates the oil into products such as gasoline, diesel oil, grease, asphalt, aviation fuel, and heating oil.

Duties: Study the physical dynamics of Earth

Education: Degree in geology or geophysics

Interests: The physical sciences, rocks, and minerals

Navigate to the Consumer Energy Information Factsheet: www.t12.lanl.gov/home/lawis/ NMNWSE/EYH/CareersBook/ C14Geolo.html for information on related careers.

Careers in Focus

Geologists use their knowledge of Earth's features to locate water, mineral, and energy resources. They use advanced scientific instruments to study the composition and processes of Earth. Geologists may spend a large part of their time in the field examining rocks, studying information collected by remote sensing instruments, conducting geological surveys, and constructing field maps.

Petroleum geologists work in the oil and gas industry. They analyze and interpret satellite and seismic data. This data helps them identify potential new mineral, oil, or gas deposits. Seismic technology is an important exploration tool to petroleum geologists. Seismic waves produce three-dimensional pictures of underground rock formations. Geologists interpret the data and identify potential oil and gas fields. This information helps reduce the economic risks associated with drilling in previously unexplored areas.

Most petroleum geologists divide their time between fieldwork and office or laboratory work. Sometimes, petroleum geologists work in foreign countries, in remote areas. They may travel to remote field sites by helicopter.

Entry-level jobs require a bachelor's degree in geology or geophysics. People holding a master's degree in geology or geophysics will have more opportunities in the workplace. Geologists often begin their careers in field exploration, as research or laboratory assistants, or in an oil company. As they gain experience, they may move into positions such as project leader, program manager, or other management and research positions.

The Age of Natural Gas

The Chinese drilled deep holes to tap natural gas more than 2,000 years ago. The ancient Persians associated flames from natural gas with Zoroastrian fire worship. Europeans, though, did not know about coal gas as an energy source until the 17th century. They distributed this "manufactured" coal gas to houses through thin lead pipes. People used it for lighting and cooking. Town gasworks became a common feature in large cities. Natural gas was not widely used until the mid-20th century.

Natural gas is a normal by-product of oil drilling. It is often found directly above oil deposits. Natural gas used to be regarded

more as a bother than as a usable fuel. It was routinely vented or burned off at the wellhead because it could not easily be put to profitable use. The exception was in gas fields that were close to populated areas, as in Europe and the United States.

In the 1970s, the world suffered oil shortages. As a result, oil prices rose steeply. People were also becoming more concerned about the environmental costs of using coal and oil. These factors forced

Natural gas gives off few emissions other than carbon dioxide. It is also a high-energy fuel.

oil producers to look more closely at natural gas. Coal gas gives off several greenhouse gases when it burns. Natural gas gives off few emissions other than carbon dioxide. It is also a high-energy fuel.

In 1950, natural gas made up about 10 percent of the world's primary energy supply. By the mid-1990s, it made up more than 20 percent, and the percentage was still growing. The major oil companies have global gas businesses. They closely link their natural gas operations to their oil operations. Today, the world's biggest gas company is Gazprom.

Natural gas is not always found with oil or coal. It exists in many places, above and below ground. Several natural processes produce natural gas. Rotting plants and animals create methane. This gas normally escapes into the air, but it can be captured from landfills, farms, and even household sewage. India now has 2.5 million facilities that trap this **biogas**. Some methane exists in large amounts underground and in ocean sediments. Deposits of methane can form much deeper below ground than oil can. Natural gas is already being removed from deposits that are 3.7 miles (6 km) underground.

 **Once natural gas has been processed and sent to a gas plant, such as the one pictured here, the nonhydrogen compounds are extracted.**

RUSSIAN RESERVES

If gas is to become the main fossil fuel of the 21st century, the Russian Federation is in a strong position to benefit. Russia has 33 percent of proved recoverable reserves—more than twice the known resources of any other country. Russia already supplies most of continental Europe with gas. Although progress on the proposed 2,500-mile (4,000 km) Yamal pipeline has been slow, it seems likely to be completed within the next few years. Other pipelines are being built to take gas through the Balkans to Turkey via Bulgaria, Macedonia, and Greece. Supplies to Serbia and Albania could follow soon. One of the most significant recent deals is an agreement between Russia and Turkey for the supply of gas using a new pipeline to be built across the Black Sea. The agreement is known as the Blue Stream project.

In the past, most efforts to find fossil fuels focused on oil. This means that the oil companies had most of the exploration expertise. Natural gas is now gaining in importance. As a result, greater efforts are being made to locate

Spheres resist pressure and are used to store natural gas.

deposits of gas that are not linked with oil, and the pace of new discoveries is increasing. It now seems that recoverable gas exists in much larger amounts than had previously been thought. Major gas fields have been discovered on every continent except Antarctica. Russia has about one-third of the world's proved reserves.

In most cases, it is not profitable to move natural gas by truck. The only efficient way to move gas is by pipeline. Pipelines that can carry natural gas over long distances first appeared in the late 1920s. It was not until after World War II that they came into widespread use. Today, gas pipelines crisscross Europe and North America. The

network in Latin America is also growing. The pipeline between Bolivia and Brazil, for example, will span 1,863 miles (3,000 km) when complete.

In recent years, major new pipelines have been built in Eurasia. These link the gas fields of Russia and Central Asia with markets in Europe and Turkey. Russia now supplies almost two-thirds of Turkey's gas and one-quarter of the gas used in Western Europe. The world's longest pipelines connect Siberia with Eastern Europe and Germany. These cover a distance of more than 3,100 miles (5,000 km). Studies are currently underway on a huge East Asian pipeline network that would run overland and underwater. It would connect the gas fields of Russia and Central Asia with the major markets of Japan, Korea, and China. It might even connect to markets in Indonesia and Malaysia.

Natural gas is changed into a liquid to cross the ocean. Liquid natural gas (LNG) is more expensive than natural gas. It is also more difficult to handle, so specially designed tankers carry it. LNG accounts for nearly one-quarter of international trade in gas. Japan is the world's biggest buyer of LNG. It gets LNG mainly from Indonesia, Malaysia, Australia, and the Middle East.

As a fuel, natural gas offers both benefits and problems. A major benefit could be its use as a bridge between economies. It can help us make the transition from a fossil-fuel economy to the hydrogen economy that is often predicted for the future. On the down side, methane is a powerful greenhouse gas. In fact, it is far more harmful to the environment than carbon dioxide. We must monitor pipeline leaks carefully.

GAS, THE NATURAL SOLUTION?

Natural gas is expected to play an increasingly large role in meeting energy demands, particularly in Asia. In 1999, about 10 percent of the total primary energy used in Asia was in the form of natural gas. This number is considerably lower than the world average of 23 percent. This difference suggests tremendous room for growth in Asian natural gas consumption. Also, techological advancements have made it more affordable to process and transport liquefied natural gas. As a result, natural gas consumption will increase significantly in Asian countries.

KEY CONCEPTS

Liquid natural gas (LNG) This fossil fuel is made by cooling and condensing natural gas into a liquid. The gas must be cooled to −260 °F (−162 °C) to make this happen. Once in a liquid state, natural gas occupies 1/600th of its original area. Converting natural gas to liquid allows it to be transported efficiently via pipeline, ship, or tanker truck.

Cleaner fuel Natural gas is considered to be a cleaner fuel. It releases fewer pollutants than oil or coal. Unlike other fossil fuels, virtually no ash is left after burning natural gas. As a result, the environmental costs of using natural gas are relatively low when compared to other fossil fuels.

Mapping Energy Resources

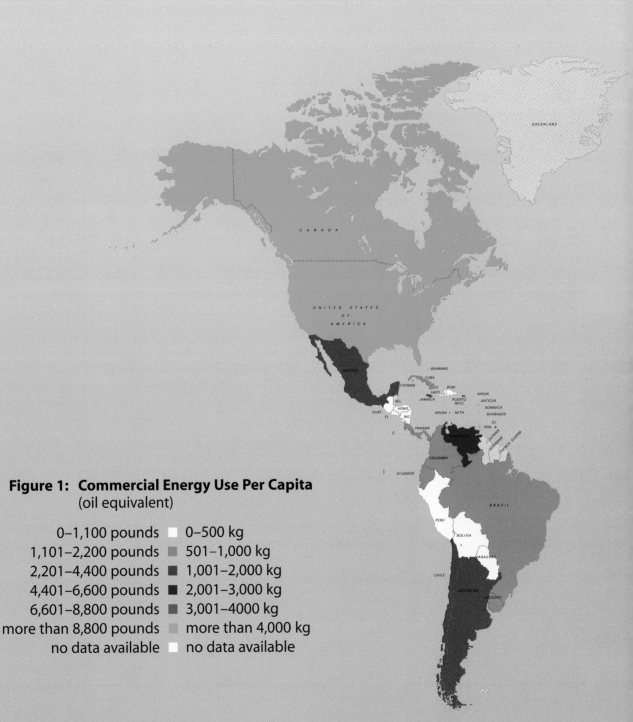

Figure 1: Commercial Energy Use Per Capita
(oil equivalent)

0–1,100 pounds	0–500 kg
1,101–2,200 pounds	501–1,000 kg
2,201–4,400 pounds	1,001–2,000 kg
4,401–6,600 pounds	2,001–3,000 kg
6,601–8,800 pounds	3,001–4000 kg
more than 8,800 pounds	more than 4,000 kg
no data available	no data available

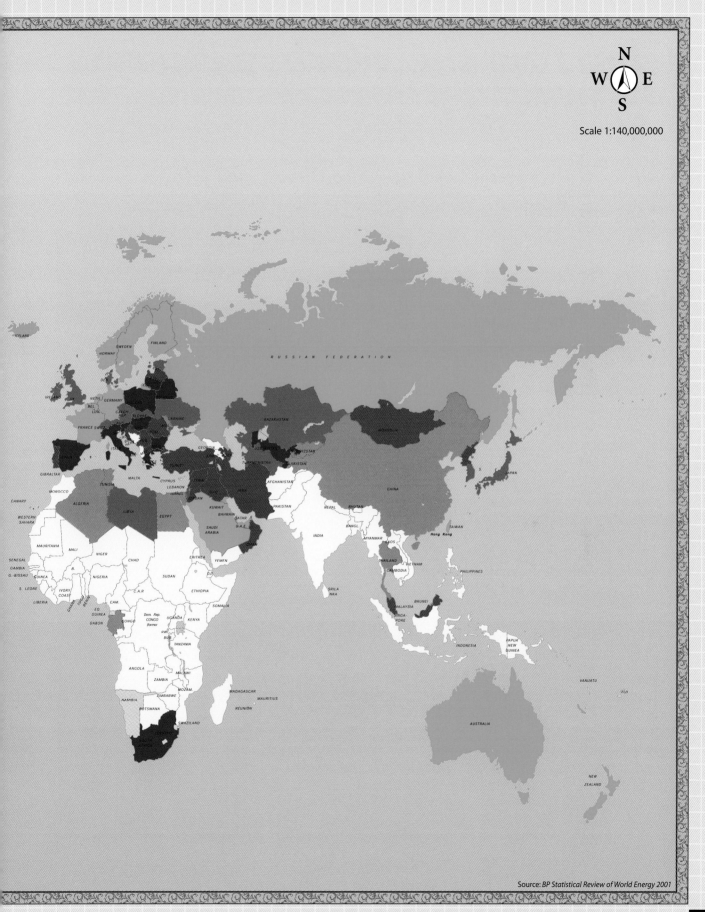

Source: *BP Statistical Review of World Energy 2001*

Charting the World's Energy

Figure 2:
World Total Primary
Energy Supply* (1999)
(* excluding wood, peat, animal dung, etc.)

- ▨ Nuclear
- ▢ Oil
- ■ Natural gas
- ▨ Hydro
- ▢ Coal
- ▢ Geothermal

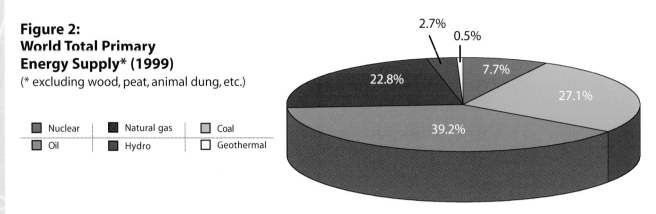

2.7%
0.5%
7.7%
27.1%
22.8%
39.2%

Total production: 9,800 million tons of oil equivalent

Figure 3:
World Coal Consumption (1950–2010)

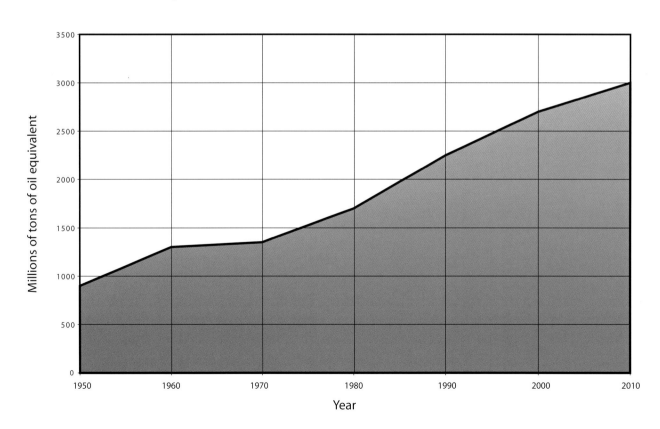

Millions of tons of oil equivalent

Year

Figure 4:
Energy Consumption (1997)
(millions of tons of oil equivalent)

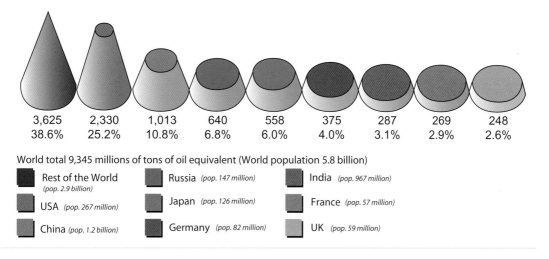

| 3,625 | 2,330 | 1,013 | 640 | 558 | 375 | 287 | 269 | 248 |
| 38.6% | 25.2% | 10.8% | 6.8% | 6.0% | 4.0% | 3.1% | 2.9% | 2.6% |

World total 9,345 millions of tons of oil equivalent (World population 5.8 billion)

- Rest of the World *(pop. 2.9 billion)*
- USA *(pop. 267 million)*
- China *(pop. 1.2 billion)*
- Russia *(pop. 147 million)*
- Japan *(pop. 126 million)*
- Germany *(pop. 82 million)*
- India *(pop. 967 million)*
- France *(pop. 57 million)*
- UK *(pop. 59 million)*

Figure 5:
World Oil Production and Consumption (1998)

This graph shows that North America, Europe, and Asia Pacific consume more oil than they produce and therefore have to import oil. The Middle East produces more than it consumes and therefore can afford to export oil.

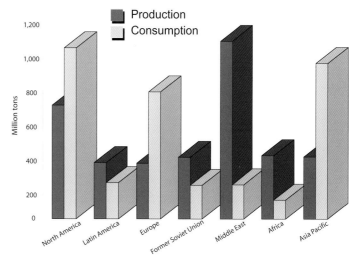

Figure 6:
Carbon Dioxide Emissions of Electric Power Plants
(tons per gigawatt/hour output)

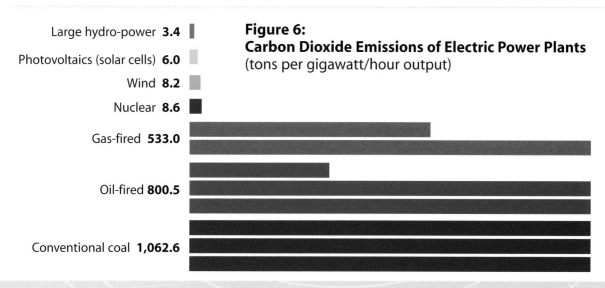

Large hydro-power **3.4**
Photovoltaics (solar cells) **6.0**
Wind **8.2**
Nuclear **8.6**
Gas-fired **533.0**
Oil-fired **800.5**
Conventional coal **1,062.6**

Fossil Fuel Technology

Concern about climate change has led to the birth of new fossil fuel technologies in recent years. New kinds of gasolines now reduce emissions. Combined cycle gas turbines increase power plant efficiency. We can also turn natural gas into liquid, clean-burning fuels.

It is even possible to take clean-burning liquid fuel directly from coal seams. Special drilling techniques turn the coal to liquid while it is still underground. This method is less harmful to the environment than first mining the coal and then turning it to liquid above ground.

In many power plants, two-thirds of input energy turns to heat and escapes. This energy is completely wasted. More efficient plants use natural gas to power a turbine. The excess heat from the turbine drives a second steam turbine. This process is called combined-cycle technology. It can increase the energy efficiency of power plants by more than 50 percent.

Combined-cycle gas systems are relatively clean power sources. They give off almost no sulfur dioxide, one of the pollutants that causes acid rain.

These systems also give off few particulates or nitrous oxides. Particulates are tiny particles that cause breathing problems and even cancer. Nitrous oxides are greenhouse gases that create smog. Gas-fired power stations also give off far less carbon dioxide than coal-fired power stations or oil-fueled plants.

Combined-cycle gas plants are only half as expensive to build as coal power stations. They can also be built much more quickly. Like all fossil fuels, though, gas is less friendly to the environment than renewable energy sources, such as wind and solar power.

One promising new technology is fluidized bed combustion (FBC). FBC works with a wide variety of fuels. This is a major advantage. FBC was originally designed for coal, but it also works with sewage sludge and **biomass**. For example, one fluidized bed power plant uses chicken manure for fuel. It is being built in the United Kingdom and will be connected to the national electricity grid. Such innovations are still the exception rather than the rule, though. Most electricity worldwide is produced by standard coal, oil, or gas power plants.

Motor vehicles use more than half of all the oil consumed every year. This is why some of

About 90 percent of the coal produced in the United States is sold to electric power plants for electricity generation.

the biggest changes taking place are in motor vehicle technology. In the 1970s, the price of oil became very high. People were looking for ways to cut back on the amount of oil they had to buy. As a result, major gains were

In many power plants, two-thirds of input energy turns to heat and is wasted.

made in the fuel efficiency of engines. Since then, concern over harmful emissions has brought about further advances. For example, the engine in one of Mitsubishi's cars can run 46 miles (74 km) on a single gallon (3.8 L) of gasoline. Toyota claims 85 miles per gallon (137 kpL) for its Prius model. This car has an engine that runs on both gasoline and electricity. Audi is

making a similar car, and various U.S. companies are competing to bring out highly efficient models that produce no emissions at all. Natural gas powers a growing number of heavy vehicles, such as trucks. The use of gas in the passenger car market is growing, too, but slowly. One problem is the scarcity of places where drivers can fill their cars with natural gas. Liquefied petroleum gas (LPG) is another possible way forward. Both LPG and LNG are cleaner than gasoline.

Many Brazilian cars run on sugarcane ethanol. Also, Iran is encouraging the use of natural gas in vehicles. In China, India, and other fast-growing markets, traditional fuels continue to be used. The cost of replacing them is simply not affordable.

Many experts say that hydrogen is the very best fuel. Hydrogen presents a few problems, though. Separating it

PETROLEUM IN PERSPECTIVE

Earth is about 4.6 billion years old. The fossil fuels we use developed about 280 to 345 million years ago. If the history of life on Earth were condensed into one calendar year, fossil fuels would have developed in late August and early September. The dinosaurs showed up in mid-September and were gone by mid-November. The earliest ancestors of humans appeared in late December. It was only seven minutes ago (about 10,000 years ago) that farmers first domesticated animals and planted crops. Australia, North America, and Africa were colonized by Europeans seven seconds ago. Widespread use of fossil fuels began about three seconds ago, and, according to estimates by World Watch Institute, U.S. Geological Survey, and the American Petroleum Institute, we have about three seconds left before fossil fuels run out.

COMMON CHEMICALS

Fossil fuels provide more than just energy. They are used to produce chemical fertilizers and some medicines, too. Fossil fuels are also the source of a wide array of useful substances we call petrochemicals. About 10 percent of the world's oil is set aside to make petrochemicals.

Petrochemicals show up in many everyday products. Here are just a few examples:

adhesives • antifreeze • antiseptics • bandages • bathtubs • candles • cleaning fluids and detergents • dyes • explosives • insecticides • makeup • paints • perfumes • plastics (e.g., compact disks, suitcases, telephones, and many forms of packaging) • printing inks • synthetic fibers (e.g., nylon stockings and carpets) • synthetic rubbers (e.g., tires)

from hydrocarbons is expensive. The process also takes a great deal of energy. One day, it may be possible to separate hydrogen from water using inexpensive solar electricity or **fuel cells**. A few companies are investing a great deal of money in solar-cell technology. Other companies are working on fuel-cell engines. These engines make hydrogen from gasoline as they run. The goal is to cut carbon dioxide emissions from engines in half.

Should the world be looking for new sources of fossil fuel? Should it be searching for ways to obtain marketable clean fuels from coal or tar sands? Many energy experts say no. Instead, they feel we should focus our attention on raising energy efficiency. Simply cutting fossil fuel waste could slash emissions by 30 percent without harming the economy.

Many rivals exist within the energy business. Publicly owned companies once supplied all electricity. Today, private companies are entering the market. These new firms offer a choice of power options. They sell fossil fuels and other power sources, too. The market is growing for energy obtained from wind, sunshine, and biomass. These sources could provide half the global energy supply by 2050. Solar cells may even be able to compete with power generators that run on fossil fuels within 10 years. Oil companies want to ensure they are ready if this happens.

An oil recovery plant or production battery separates gas and water from oil. The clean oil can then be sent to storage tanks such as these or to market.

In hydroelectric dams, huge quantities of moving water power turbines, which create electricity for consumers.

During the last century, the large oil companies have controlled energy technology. In fact, some people blame them for blocking new technologies. They claim that the oil companies have been buying new inventions that could threaten oil's lead position. It is unclear what the leading energy technologies of the future will be and who will control them.

The new, clean coal technologies also have great potential. This is especially the case in rich countries that have large coal reserves. Modern coal-fired plants can cut emissions (but not carbon dioxide) by up to 90 percent. Such plants are more expensive, though. They are not yet in common use around the world. Clean coal technology is a luxury few can afford.

The oil industry spends relatively little money on research into cleaner oil and gas.

In 1997, for example, it spent less than one percent of sales revenue on research and development. By comparison, the engineering and electronics industries spent four to six percent. Progress has been made, though. Some oil companies have created cleaner gasolines. They do this by adding fuel made from plants to their existing gasoline. People often call the mixture "green gasoline."

KEY CONCEPTS

Fluidized bed combustion (FBC) This method of burning utilizes many different types of fuels. A stream of air blows upward through fuel in a container. This suspends the fuel particles in mid-air, and they behave like molecules in a fluid. FBC is a cleaner process than simply burning coal. It traps harmful sulfur and gives off few nitrous oxides.

Alternative energy As we move toward a post-petroleum world, new energy technologies are gathering attention from countries and corporations.

Solar energy, hydroelectricity, and geothermal power are all viable alternatives to fossil fuels. Nature has great potential in terms of energy—and a practically unlimited supply. Harnessing the power of nature is perhaps our best hope for energy in the 21st century.

Pipeline Engineer

Duties: Design pipelines for petroleum transfer
Education: Engineering degree
Interests: Problem solving, physics, math, and drawing

Navigate to the Petroleum Institute Web site: 212.78.70.142/index.cfm?%20 PageID=48 for information on related careers.

Careers in Focus

Pipeline engineers plan the design, construction, and operation of pipelines that carry oil or gas from one location to another. Some of their time is spent in offices, designing plans for pipelines. They also make occasional trips to field sites. Sometimes pipeline engineers visit oil and gas pipeline sites that are under construction. There, they look for problems in design or construction in order to secure the pipe's stability. In the field, the pipeline engineer monitors the construction of his or her design.

The pipeline engineer attends to different tasks throughout a pipeline's construction. He or she has to first consider the location of the pipeline. Types of terrain and land ownership issues must also be addressed. The pipeline engineer is knowledgeable in pipeline maintenance and will consider aspects such as protection from corrosion. While the pipeline must be cost effective, it must also be strong and durable. Another part of the pipeline engineer's job is the proper placement of compressor stations so that the flow of oil and gas is maximized.

Pipeline engineers can work solely in the field supervising operations, in the office planning pipelines, or a combination of both. They may also monitor the construction of offshore pipelines. Pipeline engineers often work within a team.

Fossil Fuels Forever?

Proved reserves of oil, gas, and coal remain high. We are discovering new oil fields all the time. In fact, the amount of oil we are finding exceeds the amount being produced—but this comfortable situation cannot last forever.

Known reserves of crude oil will run dry in about 41 years at current production rates.

The United States was once the world's main supplier of crude oil. Now it relies on imports for half of its requirements. Many of its own wells have run dry. At some time in the 21st century, it seems likely that oil will cease to be used as a standard fuel for transport and power generation.

Although new oil fields are found regularly, discoveries of new supergiant fields are extremely rare. The biggest new find in the last 20 years was the Tengiz field in Kazakhstan. It contains about 24 billion barrels of oil.

Most new oil and gas reserves are in remote places. We have already developed the fields that are easiest to reach. Extracting energy without disturbing local peoples and **ecosystems** is a major challenge. The public is much more aware now than it once was of the damage that such development can do. As a result, oil companies assess new projects very carefully. The Camisea natural gas project in Peru, east of Lima, is a good example. Shell and Mobil are the oil companies leading the project. Their aim is to set new standards for community involvement and clean development. For example, no access roads will be built through the rainforest. All materials will come in and out of the area by river or by air. Gas from Camisea

The 800 mile long (1,300 km) Alaskan pipeline began to carry oil in 1977.

will eventually travel to Lima by pipeline.

Improved oil recovery methods will only delay the switch to non-fossil fuels. People have always assumed that tar sands, shale oils, and synthetic fuels would someday take over from oil. This option, though, no longer seems as viable as it once did. Both cost and environmental factors block the route. Some synthetic fuels, for example, give off even more carbon dioxide than coal does. Their processing can also require large amounts of water and result in the loss of roughly one-third of the energy in the original coal. Oil shales and tar sands present much the same problems.

Clean, renewable sources of energy offer a more sensible path. Evidence is mounting that use of fossil fuels has already

seriously harmed the environment. It is likely to cause more harm before a new energy source can be put into place.

The use of fossil fuels has already harmed the environment.

Fossil fuels have been vital in powering industrial progress so far. At some point, however, they

Automobile pollution accounts for two-thirds of Mexico City's air pollution.

must give way to renewable energy sources. Such a shift seems certain to happen, but it is still unclear how long the changeover will take and which forms of energy will replace oil, gas, and coal.

A serious dilemma faces the developing world. Is the easy lifestyle that fossil fuels provide

A BREATH OF FRESH AIR?

More than 1.3 billion people worldwide live in areas of heavy air pollution. Millions more have their health or quality of life affected by the emissions from motor vehicles, factory chimneys, and power stations. Chinese cities have concentrations of airborne particulates 14 times higher than in the U.S. The burning of coal and oil may also release trace amounts of **heavy metals**, such as lead, cadmium, and mercury. Energy that poisons the air we breathe is a poor mark of progress.

THE PRICE OF PROGRESS

China's biggest energy project is the Three Gorges Dam. When finished, it will provide a huge amount of electricity from hydroelectric power. Its energy output will equal that of about 30 large coal-fired power stations. Many people are not happy with China for going ahead with the project. About one million people now living upstream will have to find new homes. The dam will also flood a historic landscape. On the positive side, it will prevent the release of a great deal of carbon dioxide into the air by replacing traditional coal-fired power stations. It is hoped that future energy sources will avoid both harmful emissions and damage to indigenous peoples and the landscape.

worth the environmental problems they cause? Should a developing country with large coal resources leave them in the ground while its people cry out for electricity?

India and China have huge solar power potential. Unfortunately, solar power cannot yet compete with the quick and plentiful power provided by oil and coal. Standard power stations may pollute, but they provide large amounts of electricity. Most consumers ignore the costs to the environment and their health. Certainly, few in the West would choose to live without electric power or cars.

The energy choices that China and India make will have a major impact on the world. This is especially true if they follow the example set by developed countries. Resource supplies would be seriously strained. The health of the environment would also suffer. There are more than 700 million cars, trucks, and buses in the

world, and this number is rising rapidly. It is predicted there will be 1.4 billion more vehicles by 2015. China alone would have a billion vehicles if it followed in the footsteps of the United States. At present, there are only 8 vehicles for every 1,000 people in China. There are 767 vehicles per 1, 000 people in the United States.

Asia could find it difficult to finance clean ways to generate power. Clean fossil fuel energy means less pollution but not less carbon dioxide. New, high-tech power plants can run almost pollution-free. There is still no economic way to stop the burning of fossil fuels from releasing carbon into the air. If evidence of climate change grows, the pressure to stop using fossil fuels will grow.

The world's present energy system is quite wasteful. Standard power plants lose two-thirds of their fuel energy as waste heat. Motor vehicles are even more wasteful than that. Buildings and manufacturing

plants also waste huge amounts of energy.

Overall, the United States has become more energy efficient since 1973, when the price of oil skyrocketed. Still, the U.S. ranks low among developed countries

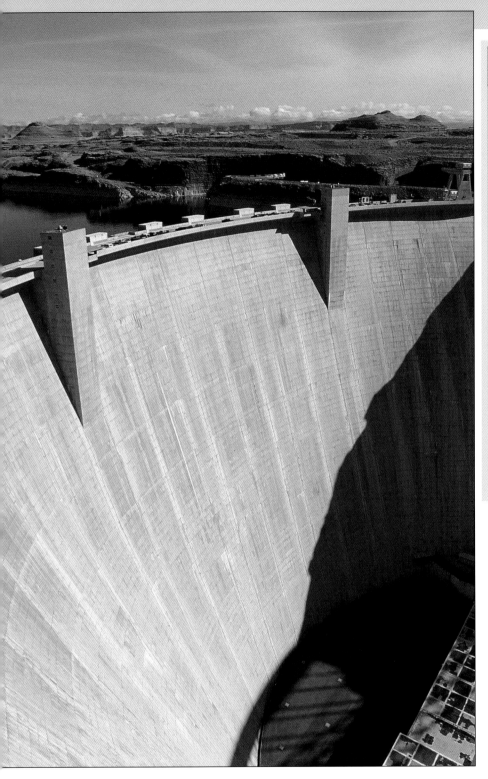

A NUCLEAR FUTURE?

Nuclear energy is one of the more viable alternatives to fossil fuels. Here are some facts about nuclear power:

1. About 20 percent of U.S. electricity comes from nuclear power.

2. The fission of 1 uranium atom produces about 10 million times more energy than the combustion of 1 atom of carbon.

3. Electricity generated from nuclear sources costs about 2.2 cents per kilowatt-hour, making it cheaper than oil or natural gas.

Arizona's Glen Canyon Dam harnesses the power of the Colorado River to produce enough electricity for 1.5 million users.

in terms of efficiency. Japan is more than twice as energy efficient as the United States. Switzerland, Denmark, France, and the United Kingdom also rank higher. On the other hand, Russia rates much lower. So do the fast-growing countries of the developing world that have large fossil-fuel resources.

Simply building more standard power stations may not be the best way to meet U.S. energy needs. Some people argue it would be better to improve energy efficiency and cut back on energy usage. Widespread use of power plants and engines that recycle waste heat could make a big difference. So could the use of variable-speed electric motors, fluorescent lamps, and building insulation. Reducing the use of electricity for heating and air conditioning could also help. Clearly, much can be done.

In 2001, the United States had 103 operating nuclear plants, accounting for 20 percent of the country's electric powe

KEY CONCEPTS

Clean development Obtaining fossil fuels without harming the environment is a growing concern. Although clean development is gaining popularity, the costs are often high. Programs exist that provide incentives for clean development.

Hydroelectric power This type of energy is generated by water at dams. Turbines at the dams change the kinetic energy of falling water into mechanical energy and then into electrical energy. Hydroelectric power is becoming a major energy alternative. *Hydro* comes from the Greek word for "water."

Duties: Design nuclear reactors

Education: A degree in science, usually nuclear engineering

Interests: Physics and engineering

Navigate to the Nuclear Energy Institute Web site: www.nei.com for information on related careers.

Careers in Focus

Nuclear engineering is concerned with the uses of nuclear processes for supplying human needs. Nuclear processes cover a wide range of technologies—from the splitting of heavy atoms, called fission, to the joining of light elements, called fusion.

Nuclear engineers research and develop the methods, instruments, and systems to harness the power of nuclear energy and radiation. They design, develop, monitor, and operate nuclear plants used to generate power. Many engineers monitor the use of nuclear fuel. They also supervise the disposal of waste produced by nuclear energy. Nuclear engineers inspect and assess nuclear power plants, including those used on nuclear-powered ships or submarines. Some nuclear engineers design power sources for spacecraft. Others develop industrial and medical uses for radioactive materials, such as equipment used to diagnose illnesses.

In the U.S., about 60 percent of nuclear engineers work in utilities, with the federal government, or for engineering consulting firms. More than half of all federally employed nuclear engineers work for the U.S. Navy.

There are some health risks involved with working with radioactive material, but excellent safety procedures exist to decrease the risks. In the U.S., there are no commercial nuclear power plants currently under construction. Still, nuclear engineers will be required to supervise and operate existing plants. Nuclear engineers will also be needed to work in defense-related areas and to improve and enforce nuclear waste management safety.

Security and Energy

The importance of the West having its own oil supplies became clear in the 1970s. This is when OPEC suddenly made the price of crude oil four times higher. The shock waves from this event caused western countries to look at other energy options. Many chose nuclear energy as a way to reduce dependence on oil. Much attention was also paid to energy conservation. Previously, oil had been so cheap that efficiency studies did not seem worth the effort.

In the early 1970s, the developed countries used most of the world's energy output. By the late 1990s, the balance had shifted. The developing countries were using almost half of the energy produced every year. Today, their energy demands continue to rise. Growth in their economy, population, and number of people living in towns drives up world energy demand.

City-dwellers tend to use far more electricity than rural people do. This holds true only if

In 1999, American energy consumers each demanded an average of 26.3 Megawatt hours.

rural dwellers follow traditional lifestyles, though. Even farms now require more energy than they once did—tractors and pesticides depend on oil.

Global energy demand is forecast to increase by more than 50 percent by 2015. In Asia alone, energy demand is likely to more than double. It could cost Asians more than $1 trillion to set up the means to distribute so much energy. China already must import 20 percent of the oil it uses. The country appears bound to use its vast coal resources if it is to meet future energy demands. This could lead to disputes between China and countries demanding global emission cutbacks.

India, too, has large coal reserves and a growing demand for energy. India would like to use natural gas from Turkmenistan. This would require a pipeline to run across Pakistan and Afghanistan, and these countries are not on friendly terms with India. Similar political problems block energy distribution in other parts of the world, too. These include the Middle East, Africa, and Latin America.

Any country that depends heavily on energy imports is bound to feel uneasy. World events could cut off the supply at any time. The United States gets one-fifth of its energy from imported fuels. This includes 50 percent of its oil. Forecasts predict that this figure will increase to 61 percent by 2015. Since 1977, the United States

has put aside oil in case of emergencies. Today, it has about 541 million barrels stored below ground in salt caverns in Texas and Louisiana. This is roughly enough oil to last the United States 53 days if all imports were to stop. Countries in Europe and East Asia generally keep much lower reserve supplies than the United States does. This puts them at greater risk because they rely on energy imports even more than the U.S. does. Most Arab countries, as well as Russia, Mexico, Nigeria, and Angola, do not need reserve supplies—they are major energy exporters.

The Middle East supplies 40 percent of world oil exports. It also has 65 percent of the world's proved oil reserves. Saudi Arabia alone has one-quarter of world

> **Global energy demand is forecast to increase by more than 50 percent by 2015.**

oil reserves. Britain, France, and the United States understood the **strategic** importance of Arab oil a long time ago. Early in oil's history, these three countries carved out spheres of influence in Iran, Iraq, Saudi Arabia, and Kuwait. Early deals favored the three great powers. Eventually, though, each Middle Eastern government took control of its own oil fields. Western oil companies still have huge investments in Middle Eastern oil and natural gas, but most of

PRICE OF OIL OVER TIME

Prices per gallon in U.S. dollars (1996 equivalent):

Year	Price
1880	$19
1900	$20
1920	$23
1940	$12
1960	$15
1980	$56
2000	$16

the oil reserves now belong to state-owned companies. The National Iranian Oil Company and Saudi Arabia's Aramco are two examples.

Money from oil sales has made the countries around the Persian Gulf extremely rich, but in recent years, even Saudi Arabia has amassed large debts. As in many Arab countries, a single powerful person rules Saudi Arabia. This state of affairs is not popular. There are strong social and political pressures for change. Arab–Israeli hostilities also complicate matters in the Middle East. Interestingly, Israel has almost no fossil fuel resources, despite its nearness to the Middle Eastern oil fields.

The United States gets 18 percent of its oil from the Persian Gulf. It is much less dependent on Middle East oil than Japan, Western Europe, or even China. Yet the United States spends billions of dollars each year to protect strategic oil supplies in the Middle East. This policy was put to the test in 1990 when Iraq invaded Kuwait. It was extremely important to the Western world that the oil fields of Kuwait and Saudi Arabia were kept out of Iraq's hands. The United States and its allies stepped in to protect the oil. They mounted a massive military objective to drive Iraqi forces out of Kuwait. Everyone knew what was at stake. There was great cooperation among countries that had different strategic interests in the region. In that case, the issue was clear-cut. A foreign invasion had to be fought off. An internal conflict, such as the overthrow of a government in a major oil-producing state, would be much harder to deal with. The oil would likely come to market, though, no matter who was in charge.

This is exactly what has happened in Iran. Years ago, nationalists banned all foreign involvement in the oil and gas industry. That policy did not benefit the Iranian people. Since then, economic reality has taken over. Iran is now keen to open its doors to foreign investment. This practical view has become the norm throughout the

> *Years ago, Iranian nationalists banned all foreign involvement in the oil and gas industry.*

world as free market forces have gathered strength.

Countries rely on one another for fuels. This fact should encourage peaceful cooperation between countries. Sadly, fights over ownership of energy resources have occurred in the past. Such energy wars could happen in the future, too. This is especially likely if oil becomes scarce and there are no affordable fuels to take its place.

China claims it owns the Spratly Islands in the South China Sea, which appear to be oil-rich. Vietnam and other countries in the region say this is not so. A similar scene is being played out in the Falkland Islands. These islands are located just off the east coast of Argentina. Recent oil exploration

OIL IMPORTS BY COUNTRY	
Israel	97%
Jordan	96%
South Korea	86%
Italy	82%
Japan	80%
Spain	70%
Germany	58%
Canada	52%
United States	48%
France	47%

KEY CONCEPTS

Energy wars Battles are often fought to protect access to energy resources. In the future, energy wars could be fought over sources of energy other than fossil fuels. For example, global warming is thought to be depleting many glaciers, yet we rely on melt water from glaciers to produce hydroelectric energy. There could be battles for the remaining rivers still fed by glaciers.

Reserve supplies Nations set aside certain amounts of energy resources, such as barrels of oil, in case of emergencies. If, for instance, OPEC decides to limit its exports, countries that do not have reserve supplies might find themselves faced with a significant energy crisis.

in the area has stirred up old arguments between Great Britain and Argentina about ownership of the islands. Which country do they belong to? Disputes also exist over seabed mineral rights. Russia, Azerbaijan, Iran, Kazakhstan, and Turkmenistan all claim rights to minerals in the Caspian Sea. Likewise, Indonesia, the East Timorese Islands, and Australia all lay claim to minerals in the Timor Sea.

In the mid-1990s, steep price increases were predicted for oil. So far, these have failed to come about. In fact, a 1998 report by the U.S. Energy Information Administration claimed that oil prices would rise only slightly over the next two decades. Other experts have less positive outlooks.

Russia and the unstable Middle East contain a great deal of the world's oil and natural gas resources. This fact is scary in a world where energy demand is rising so sharply. On the other hand, rich new sources for oil and gas have been discovered in Latin America, Africa, Central Asia, and other countries. These could ease the demand for Middle Eastern and Russian resources. The gradual shift towards renewable energy sources should also reduce the risk of future energy wars.

In February 1991, the Iraqi army ignited more than 600 oil wells, storage tanks, and refineries when it fled Kuwait.

Time Line of Events

About 125,000 years ago
Humans are commonly using fire to cook food.

5000–3500 B.C.
Ancient Egyptians harness wind energy to sail boats along the Nile River.

200–50 B.C.
The Chinese use windmills to pump water. Elsewhere, the Persians and other peoples in the Middle East use windmills to grind grain.

A.D. 100
Italian historian Pliny the Younger builds a solar home using glass, which keeps warm air in and cold air out.

1300s
The Anasazi of North America use south-facing cliff dwellings that maximize sunlight in winter.

1695
George Buffon concentrates sunlight using mirrors to melt lead and ignite wood.

1700s
Europeans begin to use coal as an energy source on a regular basis.

1800
The electric battery is invented in Italy by Alessandro Volta.

1816
The Gas Light Company of Baltimore becomes the first energy corporation in the U.S.

1821
Michael Faraday invents the electric motor.

1859
Edwin L. Drake drills the first U.S. oil well in northwestern Pennsylvania.

1870
The Standard Oil Company is formed in the U.S.

1872
The gas turbine is patented in Germany.

1882
The first hydroelectric plant is built in Appleton, Wisconsin.

1889
The first automobile is developed in Germany by Karl Benz and Wilhelm Maybach.

1890
Danish windmills generate electricity for the first time.

1903
The first gas turbine appears in France.

1910
Oil is discovered in Mexico.

1912
The British Navy begins to use oil instead of coal.

1930
The jet engine is patented in Britain.

France's first experimental solar plant, pictured here, was built in the Pyrenees Mountains.

1933
The Tennessee Valley Authority (TVA) is established.

1938
Oil is discovered in Kuwait and Saudi Arabia.

1952
The world's first nuclear reactor for commercial power is built in Pennsylvania.

1954
The Atomic Energy Act is passed.

1960
The Organization of Petroleum Exporting Countries (OPEC) is established.

1970
The U.S. Environmental Protection Agency is formed.

1973
OPEC temporarily stops exporting oil to the West. Oil prices quadruple.

1986
A major accident occurs at the Chernobyl nuclear power plant in the former USSR.

1991
The Gulf War highlights the vulnerability of the world's oil supplies.

1994
Construction begins on the Three Gorges Dam in China.

1997–2001
Eighty-four countries sign the Kyoto Protocol. The United States does not join in the agreement, believing that it will be too costly. Instead, the U.S. drafts an alternate plan to deal with emissions.

Concept Web

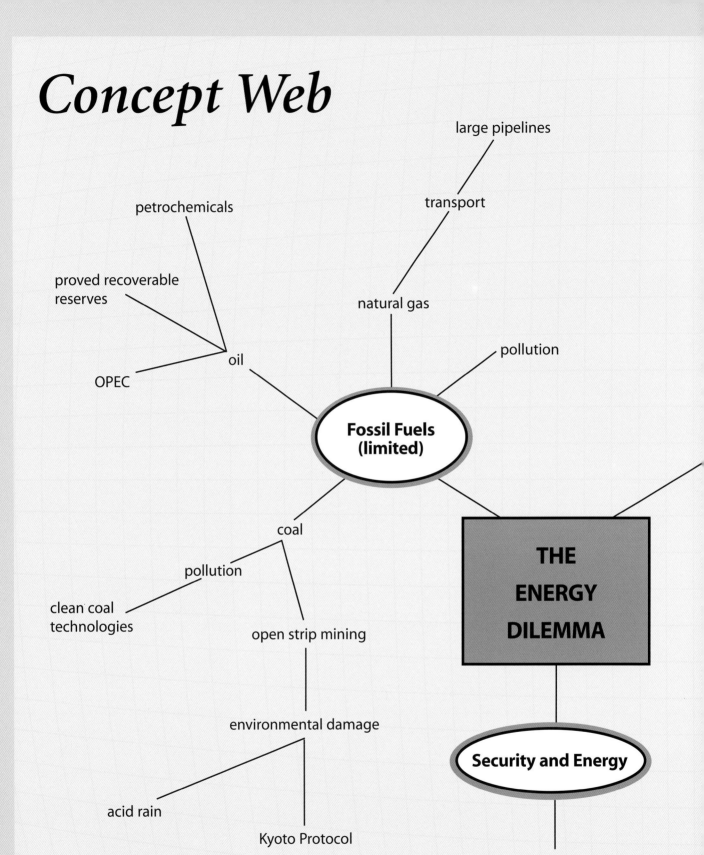

large pipelines

petrochemicals

transport

proved recoverable
reserves

natural gas

oil

pollution

OPEC

**Fossil Fuels
(limited)**

coal

pollution

**THE
ENERGY
DILEMMA**

clean coal
technologies

open strip mining

environmental damage

Security and Energy

acid rain

Kyoto Protocol

energy wars

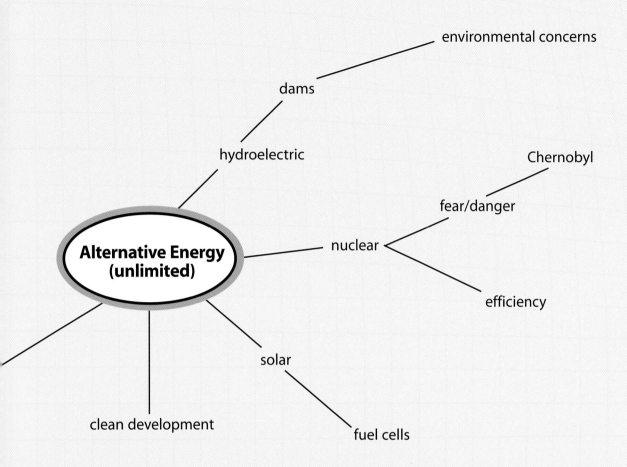

MAKE YOUR OWN CONCEPT WEB

A concept web is a useful summary tool. It can also be used to plan your research or help you write an essay or report. To make your own concept map, follow the steps below:

- You will need a large piece of unlined paper and a pencil.
- First, read through your source material, such as *The Energy Dilemma* in the Understanding Global Issues series.
- Write the main idea, or concept, in large letters in the center of the page.
- On a sheet of lined paper, jot down all words, phrases, or lists that you know are connected with the concept. Try to do this from memory.
- Look at your list. Can you group your words and phrases in certain topics or themes? Connect the different topics with lines to the center, or to other "branches."
- Critique your concept web. Ask questions about the material on your concept web: Does it all make sense? Are all the links shown? Could there be other ways of looking at it? Is anything missing?
- What more do you need to find out? Develop questions for those areas you are still unsure about or where information is missing. Use these questions as a basis for further research.

Quiz

True or False

1. Asian coal usage is expected to double between 2002 and 2020.

2. The world produces about 70 million barrels of oil per day.

3. Nuclear reactors are currently being built in the U.S.

4. Edwin Drake drilled the first U.S oil well.

5. The original members of OPEC were Canada, Iran, Iraq, and the U.S.

6. Proved oil reserves will run out in about 15 years.

7. Natural gas is liquefied before it is transported overseas.

8. More than 1.3 billion people live in areas of heavy air pollution.

9. There are more than 10 billion cars, trucks, and buses in Asia.

10. The U.S. imports all of its oil.

Multiple Choice

1. Which of the following products come from petrochemicals?
   a) makeup
   b) explosives
   c) adhesives
   d) all of the above

2. On which continents have major gas fields been discovered?
   a) North America and Antarctica
   b) every continent except Antarctica
   c) Antarctica and Australia
   d) South America and Antarctica

3. When was natural gas first drilled for and tapped?
   a) 20 years ago
   b) 200 years ago
   c) 2,000 years ago
   d) 20,000 years ago

4. The 11 OPEC countries produce how much of the world's oil?
   a) one-fifth
   b) more than one-third
   c) one-tenth
   d) one-eighth

5. Which of the following causes the most environmental damage?
   a) coal mining
   b) offshore oil exploration
   c) solar power generation
   d) geothermal exploration

6. Which country produces the most coal?
   a) Poland
   b) Russia
   c) South Africa
   d) China

7. How many oil fields have been discovered so far?
   a) about 41,000
   b) more than 400,000
   c) less than 100
   d) nearly 1,500

8. How much of the world's electricity is generated by coal?
   a) about 40 percent
   b) about 60 percent
   c) about 80 percent
   d) about 94 percent

Answers on page 53

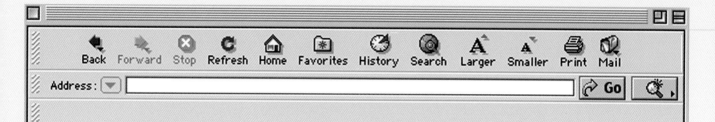

Internet Resources

The following organizations are devoted to issues and education about energy:

U.S. DEPARTMENT OF ENERGY
http://www.energy.gov

The U.S. Department of Energy's Web site has links and information related to all aspects of commercial and domestic energy use. Learn about energy production, various ongoing programs, and alternatives to traditional energy sources at this comprehensive site.

INTERNATIONAL ENERGY AGENCY
http://www.iea.org

The International Energy Agency (IEA) is a forum comprised of 26 member countries. The IEA monitors global events and was formed to address oil supply disruptions. With statistics from countries around the world and various educational links, the IEA's site is an excellent resource for petroleum-related information.

THE ENERGY FOUNDATION
http://www.energyfoundation.org

Established in 1991, the Energy Foundation supports sustainable energy initiatives. The foundation works to develop energy programs in the U.S. and around the world.

Some Web sites stay current longer than others. To find other energy Web sites, enter terms such as "coal," "natural gas," or "oil" into a search engine.

Further Reading

Brown, Paul. *Energy and Resources.* New York: Franklin Watts, 1998.

Chandler, Gary. *Alternative Energy Sources.* New York: 21st Century Books, 1996.

Fowler, Allan. *Energy from the Sun.* New York: Children's Press, 1997.

Graham, Ian. *Geothermal and Bio-energy.* Austin, TX: Raintree Steck-Vaughn, 1999.

Hansen, Michael. *Coal: How It is Found and Used.* Hillsdale, NJ: Enslow Publishers, 1990.

Holland, Gini. *Nuclear Energy.* New York: Benchmark Books, 1996.

Peacock, Graham. *Geology.* New York: Thomson Learning, 1995.

Russell, William. *Oil, Coal, and Gas.* Vero Beach, FL: Rourke Corporation, Inc., 1994.

Thomas, Larry. *Handbook of Practical Coal Geology.* New York: Wiley, 1992.

Answers

TRUE OR FALSE
1. T 2. T 3. F 4. T 5. F 6. F 7. T 8. T 9. F 10. F

MULTIPLE CHOICE
1. d) 2. b) 3. c) 4. b) 5. a) 6. d) 7. a) 8. a)

Glossary

anthracite: a form of coal that is 86–98 percent carbon and is shiny and black in appearance

biogas: a gaseous fuel produced by biological organisms

biomass: organic matter that can be converted to fuel

bituminous coal: a form of soft coal that is 45–86 percent carbon

coal oil: kerosene

coke: a substance produced by heating bituminous coal to a very high temperature in the absence of air

crude oils: unrefined petroleum

ecosystems: organisms within particular environments and their interactions

emissions: pollutants released into the air

fossil fuels: carbon-containing fuels, such as coal, oil, and natural gas, that are derived from the decomposed remains of prehistoric organisms

fuel cells: devices that produce electricity directly from chemical reactions

heavy metals: poisonous elements such as lead, cadmium, and mercury

lignite: a form of coal that is 25–35 percent carbon and is brown-black in color

metallurgical: relating to the science or art of working with metals

oil shale: hydrocarbon-rich, fine-grained rock
petrochemicals: chemicals present in fossil fuels. Petrochemicals include benzene, butylene, ethylene, propylene, and xylene.

petroleum: another term for oil. It comes from two Latin words, *petra* and *oleum*. *Petra* means "rock," and *oleum* means "oil."

proved recoverable reserves: the amount of a fossil fuel that can be removed from the ground at a profit using current methods

refined: freed from impurities

renewable energies: forms of energy obtained from natural, unlimited sources, such as the Sun, wind, and waves

strategic: essential, especially in times of war

sub-bituminous coal: a form of coal that is 35–45 percent carbon

tar sands: deposits of sand or loose sandstone that are saturated with tar.

turbines: machines that look similar to jet engines. Large blades are pushed around a shaft by a passing gas or liquid. This turns the shaft and converts kinetic energy into mechanical energy.

wellheads: the equipment that brings oil to the surface once a hole has been drilled

Index

acid rain 13, 18, 30
anthracite 11
ash 12, 25

biogas 23
biomass 31, 32

cadmium 37
carbon 11, 39
carbon dioxide 12, 18, 23, 25, 29, 31, 32, 34,
    37, 38
Chernobyl 49
coal 5, 6, 8, 10, 11, 12, 13, 14, 17, 22, 23, 25,
    28, 29, 30, 31, 32, 34, 36, 37, 38, 43, 46,
    48, 50, 51
coke 11
combined-cycle technology 30, 31

Drake, Edwin 17, 19, 46, 50

electricity 5, 6, 10, 11, 12, 13, 15, 17, 20, 31, 32,
    34, 38, 39, 40, 42, 46, 51
emissions 5, 11, 13, 23, 28, 30, 31, 32, 34, 37,
    38, 43, 47
exports 9, 18, 20, 43, 44

fertilizer 17, 32
fluidized bed combustion (FBC) 31, 34

geothermal energy 5, 28, 34
global warming 8, 9, 13, 18, 44
greenhouse gases 5, 12, 13, 18, 23, 25, 31

hydroelectric energy 5, 28, 29, 34, 38, 40, 44,
    46, 49
hydrogen 25, 32, 31

imports 9, 14, 18, 36, 42, 43, 44, 50
Industrial Revolution 6, 10

Kyoto Protocol 13, 14, 47, 48

lead 22, 37, 46

methane 13, 23, 25
mining 10, 11, 12, 13, 30, 48, 51

natural gas 5, 6, 8, 12, 17, 19, 20, 22, 23, 24,
    25, 28, 30, 31, 34, 35, 36, 37, 39, 43, 44,
    45, 48, 50
nitrous oxide 13, 18, 31, 34
nuclear energy 8, 12, 17, 28, 29, 39, 40, 41, 42,
    47, 49, 50

oil 5, 6, 7, 8, 9, 10, 12, 17, 18, 19, 20, 21, 22, 23,
    24, 25, 28, 29, 31, 32, 34, 35, 36, 37, 38, 39,
    42, 43, 44, 45, 46, 47, 48, 50, 51
oil fields 8, 9, 20, 21, 36, 43, 44, 51
oil shale 7, 8, 37
OPEC 18, 19, 20, 42, 44, 47, 48, 50, 51
open strip mining 11, 13

particulates 31, 37
pesticides 43
petrochemicals 32, 48, 50
plastics 17, 32
pollution 5, 8, 9, 12, 13, 17, 25, 30, 31, 37, 38,
    48, 50
Prudhoe Bay 20

renewable energy 5, 8, 9, 17, 31, 37, 45

smog 18, 31
solar energy 5, 31, 32, 34, 38, 46, 49, 51
sulfur dioxide 18, 30

Three Gorges Dam 38, 47
turbines 15, 30, 34, 38, 40, 46

volatile organic compounds (VOCs) 18

wellheads 17, 23
wood 10, 28, 46

Photo Credits

Cover: Oil Pumpjacks (**Comstock**); **Title page**: **Victor Englebert**; **Lorraine Bellerive**: page 21; **Comstock**: pages 4, 7, 13, 20, 34, 40, 42; **Corbis**: pages 8, 9, 15, 36; **CORBIS/MAGMA**: pages 10, 14, 24, 41; **Digital Vision**: pages 2/3, 37, 47; **Victor Engelbert**: page 19; **EyeWire, Inc.**: page 16; **Mike Grandmaison**: pages 22, 33; **Mississippi Division of Tourism**: page 6; **Photo Agora**: pages 30 (**Marie Girard**), 39 (**Patrick M. Collins**), 45 (**Novastock**); **PhotoDisc**: page 35.